FERROCEMENT WATER TANKS

A comprehensive guide to domestic water harvesting

**Felicity Lee
& Daniel Colman**

Illustrated by Berre Daneels

Permanent Publications

For Gaia

'Not only the thirsty seek the water,
the water as well seeks the thirsty

Rumi

Published by
Permanent Publications
Hyden House Ltd
The Sustainability Centre
East Meon
Hampshire
GU32 1HR
United Kingdom
Tel: 01730 776 582
          International: +44 (0)1730 776 582
Email: enquiries@permaculture.co.uk
Web: www.permanentpublications.co.uk

Distributed in North America by
Chelsea Green Publishing Company, PO Box 428, White River Junction, VT 05001, USA
www.chelseagreen.com

© 2022 Felicity Lee and Daniel Colman
The right of Felicity Lee and Daniel Colman to be identified as the authors of this
work has been asserted by them in accordance with the Copyrights, Designs and
Patents Act 1998

Illustrations by Berre Daneels

Designed by Two Plus George Limited, info@twoplusgeorge.co.uk

Printed in the UK by Bell & Bain, Thornliebank, Glasgow

All paper from FSC certified mixed sources
The Forest Stewardship Council (FSC) is a non-profit
international organisation established to promote the
responsible management of the world's forests. Products
carrying the FSC label are independently certified to assure
consumers that they come from forests that are managed to
meet the social, economic and ecological needs of present
and future generations.

British Library Cataloguing-in-Publication Data
A catalogue record for this book is available from the British Library

ISBN 978 1 85623 249 4

All rights reserved. No part of this publication may be reproduced, stored in a
retrieval system, rebound or transmitted in any form or by any means, electronic,
mechanical, photocopying, recording or otherwise, without the prior permission of
Hyden House Limited.

Contents

Chapter 1

Introduction

For most of us in more economically developed countries it is as simple as turning on a tap and out it comes. Of all the things we take for granted in the more developed world these days, clean running water is right up there at the top of the list. Little thought goes into this process for many of us and this leads to a thorough disconnect from this most precious of resources. This guide humbly aims to try to reignite the joy of clean, pure water in an attempt to make the usage and storage of it something which not only comes into consideration, but also gains an element of distinct and justified appreciation.

The current water situation worldwide leaves much to be desired; from enormous numbers of the global population still without clean, fresh, potable water, to the other end of the scale where the richest have it in such abundance that they can afford to go to the toilet in their own drinking water. The lie of the land is also in turmoil with trees, Nature's own flood defences, cut down in droves creating water run-off so extreme that from one week to the next we can oscillate from flood to drought. Flooding is becoming worse and worse year upon year, quickly leading to inundation of the already overstretched, outdated and overused municipal sewage and wastewater systems. Drought is becoming prevalent in many areas of the world, even those previously thought to be immune. These flood and drought locations are indiscriminate; Nature knows no borders, affluence, or nations and as such water scarcity will also be wholesale, but will, as always, affect those less economically well off first. Taking into account all these current happenings, where water is coursing across our fields and out to sea, alongside projections for the future of our planet where more tropical flash flooding and longer periods of drought are likely, for us it leaves a basic question hanging in the air: how can we more effectively store clean water and foster resilience?

In an effort to address this question as simply as possible, this book aims to give the lovely reader the knowledge as well as the empowerment to construct a domestic sized water tank of up to 8,000 litres for either home or garden use. The same technique can be applied to larger tanks, but for the purposes of domestic water security, the designs detailed over the course of these pages will more than suffice. It is hoped that the words and directions in this guide will enable confident, safe construction of water tanks by laypeople to enable the capturing and storage of rainwater in a safe, usable manner, in all but the harshest of climates. It must be noted that we are not engineers and tanks could be far more complex and sophisticated than the one that will emerge from these pages. This guide aims to provide you with the lived experience of two laypeople capturing, storing and using rainwater in a robust and resilient way, because after all, water scarcity will touch everybody regardless of knowledge base.

Lastly, we would like to introduce you to our tiger-like bundle of fun, Whisky. He is our omnipresent companion who loves helping to build the tanks. Throughout this book our kitty will be making a few appearances to provide some top tips that he has learnt along the way.

Whisky is here to lend a helping hand

Chapter 2

Our Journey

Our first knowledge of ferrocement tanks came about as a result of deep research for our off-grid straw bale house. Faced with a plot of land with absolutely nothing on it, we had the opportunity to implement all our systems from scratch. This left a choice: to join the services, including water, or to muddle through until the house was in a state where it was able to catch water, ideally with a roof. We chose the latter but it wasn't without its challenges to begin with, whereby we accessed water from the municipal tap in jerry cans for the first couple of years until the house was able to capture rainwater.

This initial period served us well: demonstrating exactly how much water we used on a weekly basis (150 litres). From this standpoint it was possible to project how much water we needed as a minimum. We doubled this figure. And then using the square meterage of the roof, worked out how much we could capture across the year. Finally, armed with the knowledge of the longest dry period (16 weeks in our current location), we then knew right down to the litre how much we needed to store. Then came the big question, what is the best way for us to store the rainwater safely, securely and on our shoe-string budget? The options were various, the Rolls Royce of storage being stainless steel, another option being underground plastic vats, and well, our ultimate choice, ferrocement tanks.

Chapter 3

Why Ferrocement?

Cost and ease

The combination of cheap, easily available materials, margin for error, suitability for amateurs and flexibility in terms of size and shape, make ferrocement tanks an ideal choice for anyone looking for long-term water security on a budget. Of course this list sounds magic, but the flip side of it is this technique is most suitable for those who are happy to get their hands dirty. The required labour time is substantial, and as such, few are seen in countries with high labour costs. With a few practical skills and a little perseverance, it is perfectly possible to achieve water security for a minimal cost. We built two tanks and achieved water security for less than the cost of joining the local water network!

Simplicity

In essence this technique involves building a cylindrical cage consisting of steel rebar and chicken wire and then hand-plastering it with a cement mortar. It lends itself to group projects with volunteers owing to the simplicity of the tasks involved and the tendency for those new to things to be over-meticulous – never a bad thing! The tools required are absolutely minimal and it can be done (as we have) with no power tools whatsoever, although you might end up with a few aching muscles and toned triceps...

At its simplest, this design provides gravity-fed water storage with no moving parts other than external taps. It can also provide a reservoir which has the possibility of creating a pressurised or non-pressurised water system which can be plumbed into a house, cabin or other building.

Cleanliness through design

The cylindrical nature of ferrocement tanks allows for a constant yet almost imperceptible movement of the water therein, through convection. By definition, the movement ensures that the water is never stagnant and therefore far less prone to the build up and harbouring of bacteria and algae, and essentially keeps the water 'alive'. The design outlined in this guide allows for cleaning to be a breeze whenever it is deemed necessary. Furthermore, it does not leach in the same way that plastic does and due to the fact that it is concrete, no sunlight can enter and spoil the water by encouraging algae build up. The only superior material would be stainless steel as previously mentioned; it is a great material for water storage but not without its drawbacks, namely finding one sufficient in size, at the right cost and with the ability to be transported.

Uses of ferrocement tanks

The uses of water storage are limitless but here are a few:

Complete off-grid water independence

A back-up supply in water-scarce or precarious regions

A supplementary supply to reduce water bills

A garden supply to reduce water bills

Water supply for a remote cabin or house

A way to reconnect with an elemental force that we take for granted

A cleaner water source for those sceptical of local water quality

Where connection to the municipal supply is impossible.

A well designed tank with a few key features can provide clean, cool, drinkable water year-round, with no electricity, filters or moving parts and absolutely minimal maintenance. The tanks need a scrub out every year which, owing to strategically placed drains, takes about 45 minutes.

Water harvesting and storage is a profoundly satisfying experience which can help recapture the sense of awe our ancestors must have felt as clouds rolled in and gave freely of this life-giving element. Taking a cool shower during a heatwave in crystal clear, cold rain water, harvested nine weeks before, is a memorable experience indeed.

Chapter 4

Principles of Water Storage and Construction

How you meet these principles can vary depending on what is applicable to your particular context:

Volume – A cylinder will hold more water relative to surface area than a cube, making it more material-efficient.

Strength – Curved surfaces are stronger than flat ones, meaning your ferrocement tank will support eight tonnes of water with a 3cm-thick wall, again saving materials and reducing cost.

Potability – Avoiding right angles by making a cylindrical tank (perhaps with a domed wine bottle-esque floor, see page 29-30) will promote water circulation, convection and aeration, meaning that water does not stagnate in corners as it would in a cube. We have had occasions when no water has entered our tank for nine weeks and we had a second heatwave within a two-week period with temperatures in excess of 40°; the tank was in full sun yet the water inside remained cool, clean and perfectly drinkable.

Flexibility and ease of use – Your tank should be appropriately positioned, sized, easily maintainable and have the necessary taps, drains and other features to ensure ease of use and minimal ongoing upkeep.

Responsibility – Becoming an active participant in the cycle of water by storing and using it automatically leads to a greater awareness of resource consumption which spills over into other areas of your life. A dry toilet for instance would be a sensible addition instead of flushing all that lovingly collected water.

Chapter 5

Tank Features

There is no limit to how complex these tanks can be, but we will stick to the absolutely necessary features according to our experience of having built three tanks and enjoyed water independence for upwards of two years:

A **drain** in the tank floor, preferably with the ground sloped towards it, although if you mess this up it may only cost you a few extra minutes of sponge rinsing once a year. Keep in mind what type of fitting you plan to place on the outside; this will determine the size and type of pipe you use. Make sure the pipe and tap, stop valve or other, are designed to go together and are of a high quality to ensure longevity.

A **tap** for getting the water out, about 15cm above the tank floor with a fitting appropriate for intended use. This 15cm allows for organic matter to settle below the point from which you intend to draw water. It might be that you would like the tap to be attached to a hose, thus a suitable hose fitting will be required on the end of the tap.

A **trough** may be required underneath the bottom tap if you don't intend to pipe it out, to allow space for something, such as a bucket, to fit under the tap to collect the water.

A **second tap** around half way up the tank for ergonomic access when the tank is over half full.

An **inlet pipe** from your roof/gutter. This may require a filtration device or mesh to prevent too much debris entering the tank.

Simple and effective 'leaf eating' barrier devices are available to buy ready-made or it is possible to create your own version. The general principle is to build a barrier using metal or plastic mesh (metal is more hardy) on top of the tank inlet that sits at a diagonal angle. The drainpipe comes down and matches this angle, but sits a couple of centimetres above it. The slope allows for any debris to fall off the mesh preventing blockage over time. A barrier designed on a flat aspect will accumulate leaves and other gunk, quickly becoming blocked. Another option or addition, if the water is coming from a roof, is to place a barrier in the top of the drainpipe, in line with the gutter. However this will require regular checks to make sure debris hasn't built up too much thereby causing the gutters to overflow. Again, these mechanisms are easy to find in hardware shops and ultimately depend on how much debris you are likely to accumulate.

An **overflow pipe** situated just below the level of the inlet pipe (although not directly under it); if it is too far below you will waste capacity. Also think about bug-proofing. This could be a metal mesh of some kind. Also worth thinking about is where the overflow water is going to go. It could just overflow and percolate into the land, or you may want to direct it towards a swale or trough. The overflow pipe itself should be cut so the end inside the tank is on a horizontal plane, preferably as close to the top of the tank's shoulder as possible, without being any higher than the bottom of the inlet pipe. This helps with the removal of floating debris and utilises the entire capacity of the tank.

A **removable lid** made of ferrocement with some kind of mechanism (hook or rings for instance) to allow two people to lift it off for cleaning.

An **additional outlet pipe** should be placed 15cm above the bottom of the tank if you plan to plumb the tank into a house or cabin. It is worth reiterating here that pipe sizes will be in accordance with your local norms. Visit the plumbing section of your local hardware shop and take a notebook. The outlet is primarily there to allow the water to be plumbed into something, for example into a house, cabin or other. Therefore, plumbing grade pipe, usually called 'polyethylene pipe', should be used to plumb it into your house. Before starting construction of the tank, check your local supply of plumbing materials and the most common sizings for pipes, as this will guide your choice of fittings for tank outlets. Consider if the pipe will meet a pump or be gravity fed and ensure you buy lengths of pipe, joints, clips and fittings accordingly. There tends to be some discrepancy between countries as they have slightly different piping and joint sizes. It is worth adding a stop valve just after it leaves the tank so you can isolate it from the house if needed. Ferrocement tanks work very well when plumbed into evacuated glass tube solar water heaters.

A **mechanism to see the level** – Optional. You can literally just feel where the tank becomes cold (as we do), install a simple floatation system or even just lift up a hose attached to the bottom tap until the water stops flowing.

Further reading: For a far more exhaustive list of possible features and types of tank applications, we highly recommend *Water Storage* by Art Ludwig.

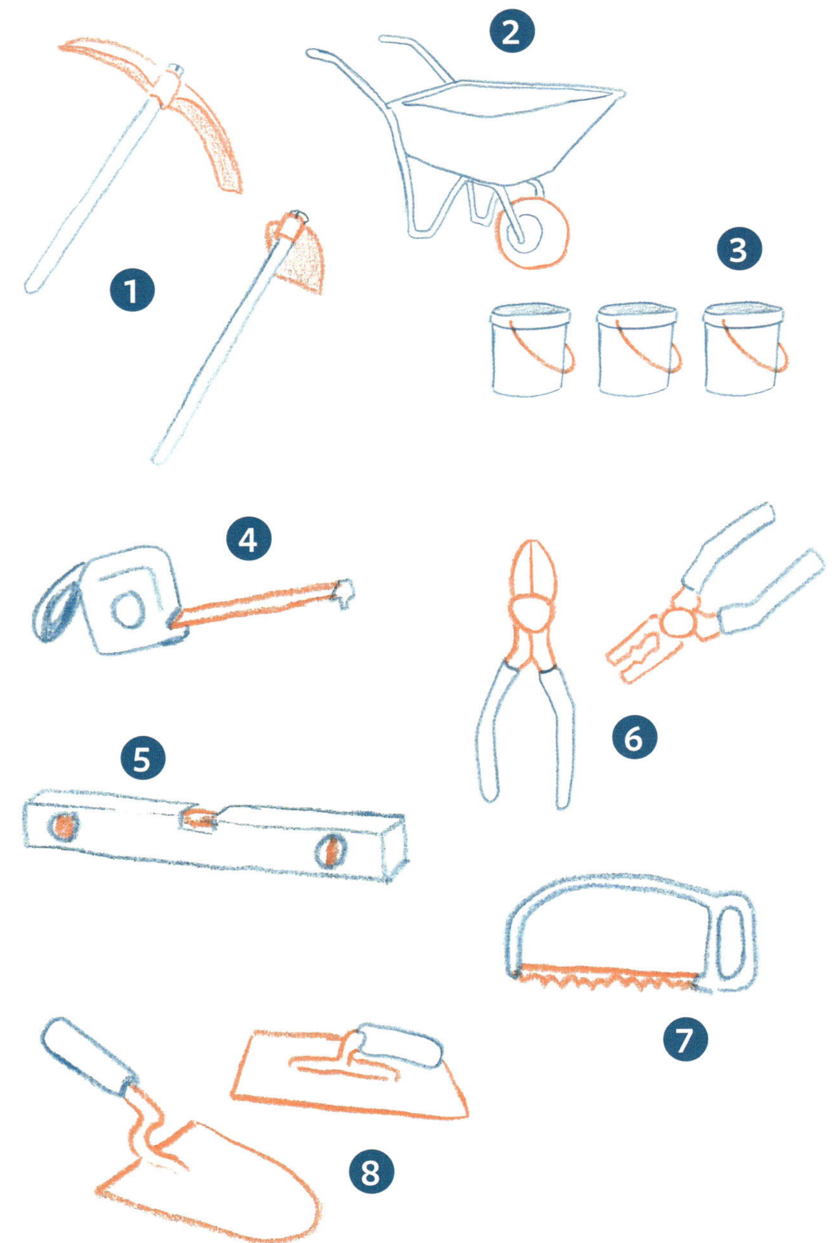

Necessary tools

A **pickaxe** or **backhoe** for digging the foundation. **1**

A **shovel** for shifting earth, sand, cement and gravel.

A **wheelbarrow**. **2**

Buckets – at least three for, well, everything! **3**

A **tape measure** for sizing the rebar and measuring the diameter of the steel hoops. **4**

A **spirit level** can help you achieve the desired slope in your foundation – but your eyes will generally do a sufficient job. **5**

Pliers / Wirecutters – At least two pairs, more if you have more hands, for twisting and cutting wire. **6**

A **hacksaw** for cutting the steel rebar. **7**

A **piece of pipe** or equivalent for bending the rebar – we used the hollow rungs of a ladder!

Trowels of various sizes, but ones with pointy ends (a diamond) are not very useful; round-edged ones and rectangular ones are best – oh, and a bucket trowel to empty out the buckets efficiently. **8**

A **large paintbrush** for applying the water-proofing cement paste mix at the end.

Optional tools that save heaps of time

A **cement mixer** is not absolutely necessary; you can mix concrete under your own power, but it will make the process far easier. ❶

A hand-pumped **pressure sprayer** is extremely useful, especially if you don't have access to running water – it is crucial for dampening, smoothing and cleaning. ❷

Hog ring pliers can be worth their weight in gold, saving you painstakingly twisting hundreds of pieces of wire. ❸

Whisky's Top Tip

Hog ring pliers will save you a lot of time and cursing.

Materials

You will need access to the following materials in varying quantities depending on the size of tank you intend to build:

Steel rebar of either 8mm for a tank up to 5,000 litres or 10mm for up to 8,000 litres (although if you can only access smaller stuff, just make more hoops!). It is nearly always rusty when you buy it, so don't worry.

Galvanised wire for securing the rebar and attaching the chicken wire, unless you use hog rings for the latter in which case you will need it only for the rebar hoops stage. This can come in various sizes; if you can only find thin stuff, say, less than 1mm thick, then just double it up on each joint. The ideal thickness is 3mm galvanised wire.

Chicken wire with the smallest gaps you can find – this will massively impact the ease of the plastering process – we can't emphasise this enough!

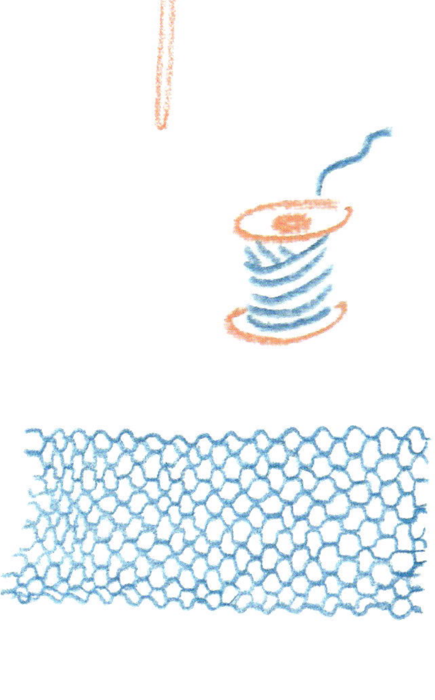

Sand from a builder's merchant, usually 2-4mm, often called 'builder's sand' or 'plastering sand'.

Gravel for the foundation mix, no bigger than 20mm.

Cement in 35kg bags.

Water for making cement and also cleaning and keeping new plaster wet – you'll need plenty, either with your own supply or ask a kind neighbour to lend you their hosepipe.

Tarpaulin(s) to protect the tank from sun or frost and to mix cement if doing it by hand.

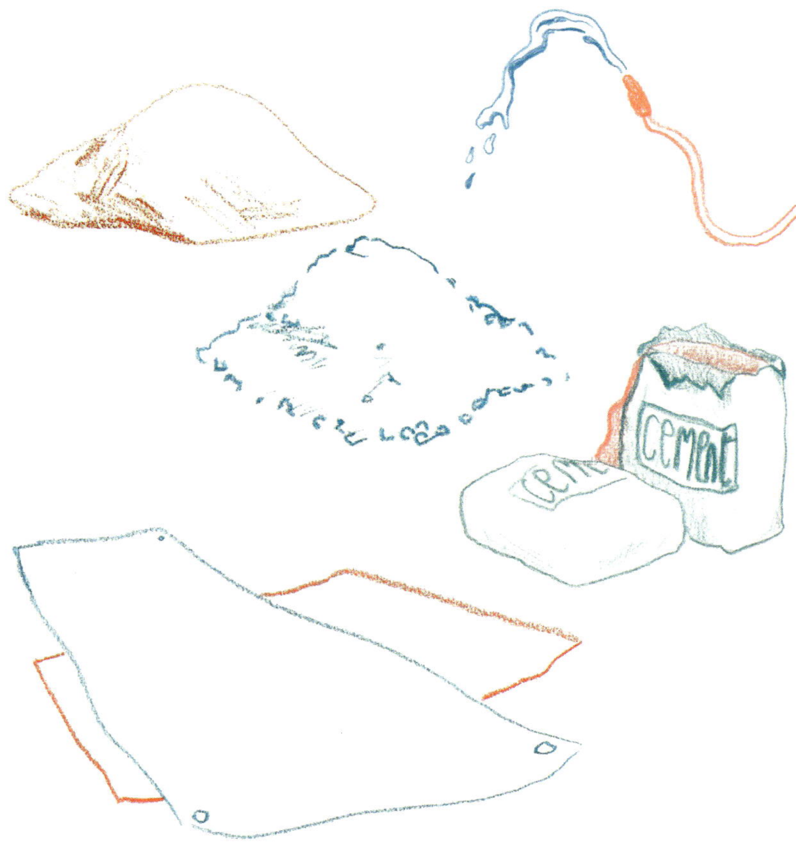

Chapter 6

Considerations

Applying the cement to the chicken wire requires a knack. The greatest success we had was having one person outside the tank with two large rectangular trowels and one person inside pressing the cement through the wire by hand.

Use chicken wire with the smallest possible gaps; this will massively increase the ease with which the cement sticks.

Hog ring pliers will save you a lot of time and reduce cursing.

Add a tap half way up, or higher perhaps, to give you more ergonomic access when the tank is full or at least above the level of the tap. This is great for watering cans in the garden and avoiding a bad back, not to mention having an idea of how full it is.

Get some help – extra hands make for lighter work, sharing knowledge and having fun.

Get the chicken wire nice and tight onto the rebar skeleton; it will make the plastering much easier. Don't settle for baggy wire joints; ensure you practise the twisting of the wires to get it really tight, see page 23.

The rebar hoops are hard to shape perfectly; shoot for perfection but accept it won't be! The strength and performance won't be affected, just the aesthetic as the tank is built. Overall, the cylindrical shape will win out by the end.

Making a domed floor (like the bottom of a wine bottle) will increase the strength of your tank and potentially reduce material usage. In reality, our domed floors don't seem to save much material, due to erring on the side of caution in terms of concrete thickness, but it definitely makes it stronger.

Rebar usually comes in 6-metre lengths. If you have a small trailer try to somehow bend them in and avoid cutting, as you will need the long lengths to make hoops. Making hoops out of two separate pieces is possible if totally necessary (for a larger tank); see page 24.

Clean tools thoroughly and immediately after work before the day is done; crusty tools, gloves and buckets are a terrible way to start the day!

To make access for plastering inside the tank easier and to avoid climbing in and out of it by dodgily going in through the top using a ladder (this only applies to tanks which require a join in the hoop), leave the section with the extra bit of hoop open and don't cover that area in chicken wire – once the inside has had its second coat of cement, you can then wire in the extra rebar, attach the chicken wire and plaster up the hole.

The two big questions

The two biggest questions at this point are:

1. Where should I put my tank?

2. 'How big should it be?' with the related question: 'What height/ diameter should it have?'

Tank placement

Ask yourself these questions and you should be on your way to an appropriately positioned tank:

- Does it need to be lower than a gutter? Always yes!

- Would you like to benefit from gravity by situating it above where you intend to use it?

- Is there a position which will allow more shade or sunlight (depending on climate) to either cool or warm the tank?

- Where do I need to use the water?

- Would you prefer it to be hidden or on display?

- Are there any tree roots which could interfere with the tank or even be damaged by the work to be carried out?

- Is there decent, solid subsoil or bedrock onto which I can place the tank?

Tank size

Our two potable tanks of 8,000 and 3,000 litres provide all our domestic water needs. We have a low consumption house with a dry toilet and have become very conscious of our usage. The size of your tank will depend on three variables:

- The longest dry period

- Average rainfall

- Your average consumption and whether or not this can be reduced through either attentiveness or some changes in your house, installing a dry toilet for instance – which will reduce your water usage by around 30%.

Making a very low, wide cylinder isn't advisable, due to losing the material efficiency of a higher tank, increasing exposure to the sun and having to build an enormous foundation. Equally a very high, narrow cylinder is unwise too, the height of the structure making it potentially unstable, hard to access for both construction and cleaning, not to mention looking a wee bit phallic.

With these observations in mind you will probably end up with a tank which is roughly head height or perhaps a touch higher and then a diameter to suit your capacity goals. There is a helpful calculation for creating a tank with the desired volume which has something to do with pies.

Working out the calculations can really take you back to school geometry and possibly give you a small headache. Due to the nature of circles, it isn't possible to easily work out the calculations of the cylinder starting with the volume. The best starting point is the radius of the tank floor. Thus, it is often best to start with a

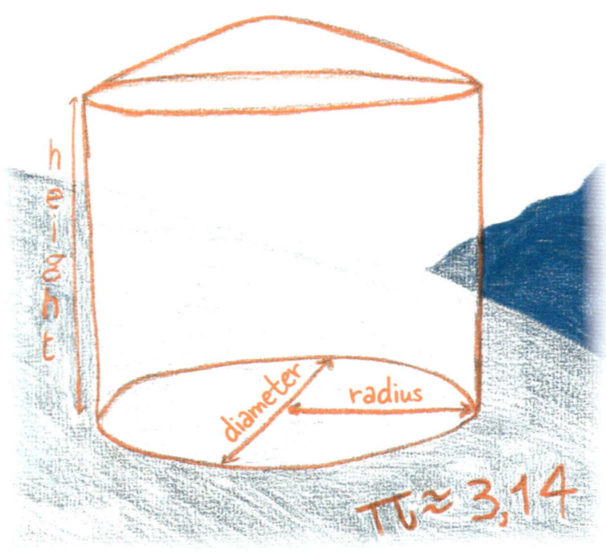

guesstimate here and keep playing with the number until you reach your target volume. If you already know your height parameter this can really help.

Let's imagine you would like to build a tank of roughly 6,000 litres that is 2 metres high to the shoulder (usable water storage space). Begin with a radius of 98cm, a diameter of 196cm. Then employ our old friend Mr Pi π. Pi multiplied by radius multiplied by radius equals the bottom surface area of the cylinder. In this case 3.14 x 98 x 98, which then needs to be divided by 10,000 to give the metres squared. This equals 3.01m², the surface area of the bottom of the tank. Now multiply 3.01 by the height, in our case 2m, and you have the tank volume which is 6.031m³, and then multiply this number by 1,000 to move it to litres, equalling 6,031. Sadly, complete and pleasing round numbers don't really make an appearance here due to Pi, which means this tank will have a handy extra 31 litres.

For those of you whose brains work better with equations rather than words, the same calculations can be found below:

$$\text{Radius (cm)} = \text{Diameter (cm)} \div 2$$

$$\text{Surface area of tank floor (square metres)} = \text{Pi x radius x radius} \div 10,000$$

$$\text{Tank volume (litres)} = \text{Surface area of tank floor (m}^2\text{) x height (m) x 1,000}$$

Put these equations into a spreadsheet if you like, then you can play around with the dimensions until you get what you want!

Chapter 7

Designing the Cage

Having decided upon the size of the tank, it is now time to design the cage. This is necessary to ensure the correct amount of rebar is purchased and so that all of the component pieces of rebar can be cut and bent prior to putting them together.

Uprights

The cage will consist of upright pieces of rebar together with horizontal hoops. The number of upright pieces and the number of hoops should be sufficient to ensure the right amount of strength relative to the tank; this comes down to spacing. As a rule, the gaps between rebar should be no more than 30cm, apart from at the bottom, where the first horizontal hoop will sit much closer to the bottom of the foundation, being only 10cm above the foundation cement adding extra strength to the critical floor/wall junction.

As an example, let's use the 6031L tank we worked out in the last section. First, it is necessary to have the circumference to hand. In this case it is 1.96m x 3.14 = 6.15m.

Diameter (cm) x Pi (3.14159) = Circumference (cm)

This is then divided by 30cm (0.3m) to give the correct spacings, which works out at a total of 20.5 uprights. For this example it will be rounded down to 20. Therefore, 20 upright pieces of rebar will be needed for this 6031L tank. At this point it is crucial to remember

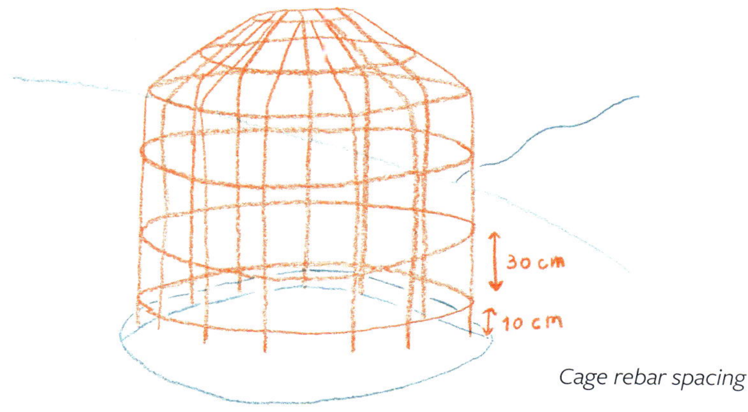

Cage rebar spacing

that the upright measurement must take into account the roof of the tank and its desired slope (somewhere between 5–15°), the shoulder point at which the uprights bend in towards each other, creating the tank roof and opening. The opening is usually 70cm in diameter to allow a human to access. Be generous in this measurement; it is far better to trim an upright that is too long than have to cut another piece because it was too short.

Hoops

Next up are the hoops. In this example, the tank height is 2m and the first hoop will be only 10cm above the foundation to give extra strength at this critical wall/foundation junction. Removing this 10cm from the calculation, the remaining height can be divided into 30cm gaps. 1.9m divided by 0.3m equals 6.3. This number indicates how many hoops will be needed for the tank. In this case, you can choose to round up or down. Generally, it is better to err on the side of caution and round up to 7; after all the material cost is very low.

Rebar usually comes in 6m lengths, depending on where in the world it is being located. For this tank, rebar with a diameter of 1cm is preferable. If the circumference of the hoops is greater than 5m (which includes an overlap at the join of 50cm from both ends, equal to 1m) then it will be necessary to extend the hoop by making it out of two pieces. Going back to the hypothetical 6031L tank, which has a circumference of 6.15m, each hoop will require one full 6m piece of rebar plus a 1.15m piece of rebar to overlap. This will give the correct circumference of 6.15m with 50cm overlaps. Overlaps need to be firmly secured with three wire ties.

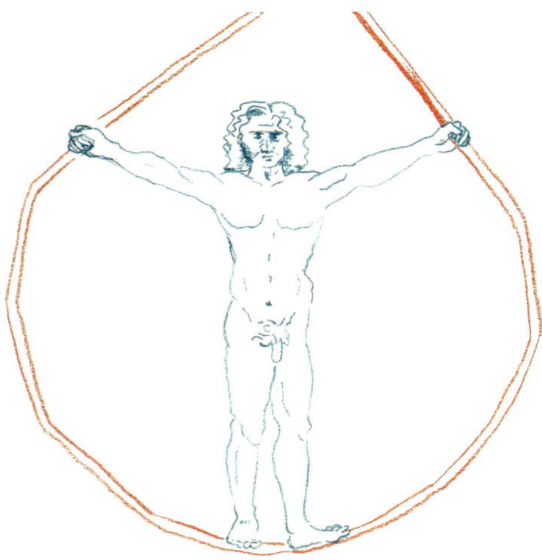

Try as far as possible to make the hoops circular. Inevitably, there will be kinks and bends, but they tend to even themselves out once the tank goes up. The larger the hoops the harder it is to keep them true, but with helping hands and your best impression of Leonardo's Vitruvian Man, the end result will be fine.

The larger hoops which require two pieces of rebar have one advantage: two or three of the hoops can be left incomplete leaving a gap in the side of the tank which can be used for access during the plastering phase. The sections needed to complete the hoops can then be added afterwards, chicken wire attached and the hole plastered in.

Rebar foundation

The rebar foundation consists of concentric circles of rebar, the outer-most of which is the same size as the other horizontal hoops. The rest become subsequently smaller, following the 30cm spacing rule.

All of the wall uprights will need to be fixed to steel protruding 50cm from the foundation once the concrete is set. So make sure the number and spacing of your right angled pieces matches up to the number and spacing of the uprights you'll need for your tank wall. Be sure to cover these sticking up pieces of rebar on the work site; they can be really quite dangerous. Anything can be used, from squares of wood to old aluminium cans. Hence, it will be necessary to make as many right angled pieces as there are uprights.

The right angled pieces will be cut to a variety of lengths along the foundation bottom. Some will reach all the way to the centre point,

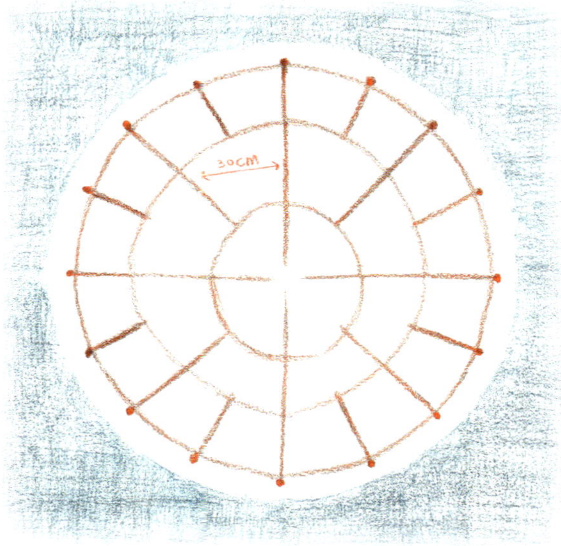

A foundation rebar design for a tank with 16 upright wall pieces

The foundation rebar design on a cone shaped excavation with protruding right angles ready to meet the uprights – be sure to cover them for safety until the uprights are attached

others will finish within the other concentric circles. As long as each piece is fastened to at least two concentric circles, all will be well.

In summary, adding up all of the various sections in this chapter should allow for a reasonable estimation of how much rebar your tank is going to need, ensuring a stress-free trip to the builder's yard.

Rebar for the trough and the lid have not been mentioned, but it is a safe assumption that there will be enough for this in the way of offcuts from the main tank structure.

Bending rebar

The possibilities are endless when it comes to bending or shaping the rebar. Body and arm strength will suffice for the large hoops. But to achieve the right angles, some mechanical help will be required. We used the hollow rungs of a ladder and also a thin metal pipe, and at one point a large plank of wood. They all need mass (body weight!) to work. There are other ways, including using two offcuts of rebar which are notched to receive the rebar, creating a lever. Or there are professional rebar benders/cutters available.

Tying the wire knots

This knot will be used at almost every stage of the tank build, so it is well worth having a few practice runs to make sure you are able to get it nice and snug around two joining pieces of rebar.

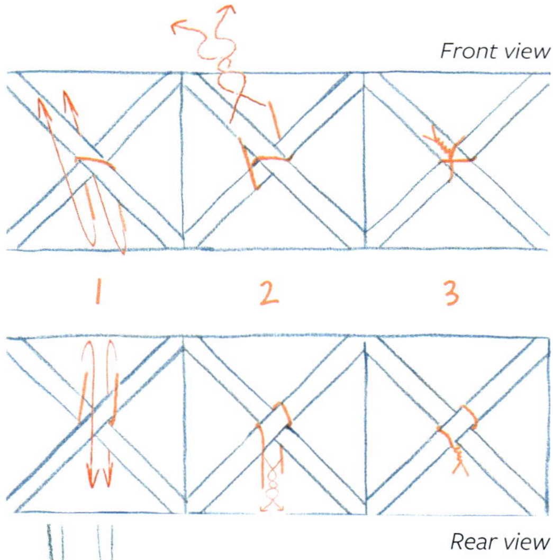

Front view

1   2   3

Rear view

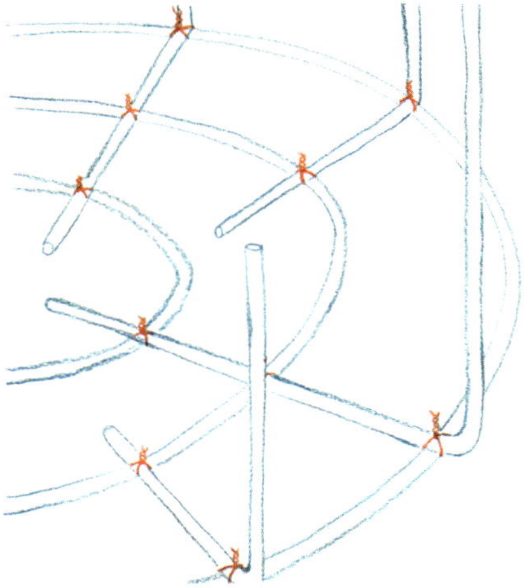

Joining pieces of rebar using galvanised wire

Chicken wire

When choosing chicken wire it pays off to spend a little more money here and buy the wire with the smallest gaps. However, make sure this is not 'square welded wire mesh' as this is weaker. Chicken wire is galvanised and has hexagon shaped holes which are twisted rather than welded.

Throughout our own tank builds we have tried with various sizes of wire and by far the best result was with the smallest holes of 15mm in diameter.

To calculate the amount of chicken wire you will need requires another fun, simple calculation. Multiply the circum-ference in metres by the height in metres. Multiply this by two to get both inside and outside, and then make sure you add enough to account for the roof as well. For our hypothetical 6031L tank, this works like this: 6.15 x 2 = 12.3. Then, 12.3 x 2 = 24.6m² of wire. In this case, buying 30m² of wire will be enough for the roof and lid too.

Whisky's Top Tip

Use chicken wire with the smallest possible gaps; this will massively increase the ease with which the cement sticks!

Chapter 8

Building the Tank

The following instructions apply to any size of tank between 3,000 and 8,000 litres.

Step 1: Make sure you have everything you need! So easily overlooked, but running out of cement on a Sunday or when you are miles away from a town is a nightmare.

Step 2: The first rule of fight club is... the second is... make DOUBLY sure you have everything you need!

OK, enough kidding around...

Step 3: Dig a foundation hole

Around 80cm bigger than your tank diameter, this will give you some working space around the tank which is crucial for plastering and stops the soil collapsing into your hole and making a mess.

Go down to solid subsoil or rock; there shouldn't be any organic material under the foundation such as roots or grass.

If you are keen, attempt to slightly shape the hole into a dome, like the bottom of a wine

A flat foundation hole down to solid subsoil

bottle, between 15–40cm high in the middle. This can be done by hammering a piece of rebar dead centre, drilling a hole in a piece of wood the length of the hole radius, placing it over the rebar and, by rotating it around the hole, using this as a rough guide as to where to excavate more/less.

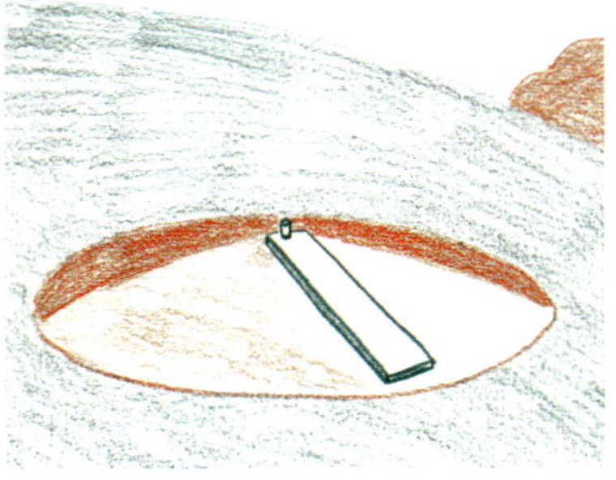
A domed foundation hole

Step 4: Steel for the foundation and preparing to pour

Make a hoop the size of your tank circumference; don't forget to include an overlap that is 50 x the rebar diameter e.g. with 10mm rebar this equates to a 50cm overlap with three wire ties holding the overlap in place. Place the ring in the centre of the hole.

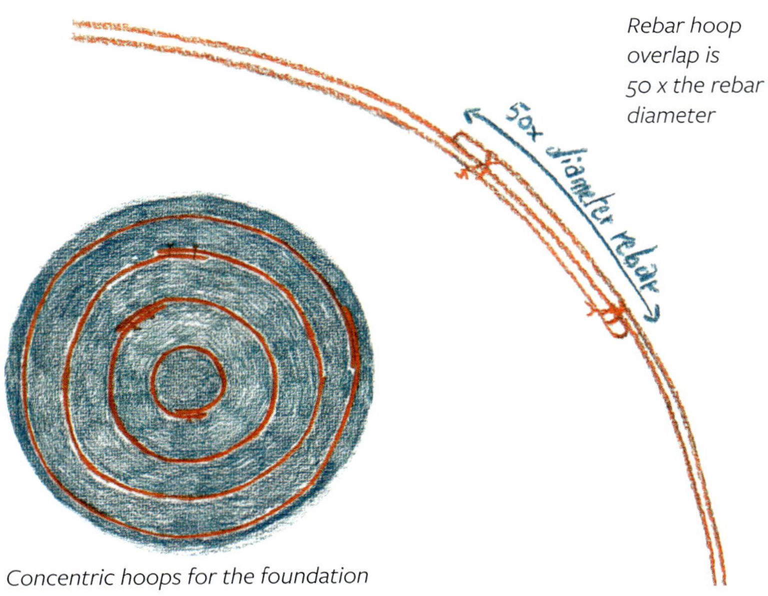

Concentric hoops for the foundation

Make and place another, smaller hoop, 30cm closer to the centre of the foundation hole, then repeat – you will end up with three or four concentric hoops evenly spaced at intervals of no more than 30cm. Note: the pressure pushing outwards from the tank is taken by the hoops on the side of the tank – the metal work in the foundation gets its strength from the depth of the cement and its proximity to steel, so the hoops don't have to be perfect.

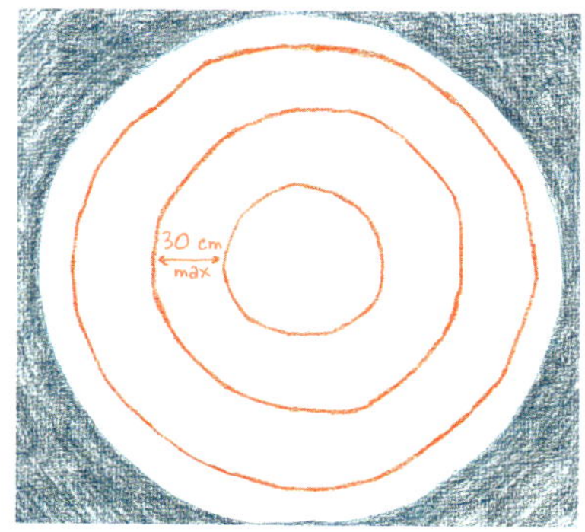
Hoop spacing for the foundation

Make as many right angle pieces as you have uprights in your walls; these will protrude from the concrete foundation once it is poured and be used to fix the uprights for the tank walls. There is no need for all these pieces to be long enough to reach the middle of the circle as it would be too crowded. Just size them appropriately, making sure the gaps between each piece of rebar are no bigger than 30cm. The smallest ones should still be fastened to at least two hoops.

Note: you'll need them to be bent slightly less than 90 degrees if you intend to have a domed floor.

As mentioned in the last chapter (page 25), if there are going to be 16 upright pieces of rebar, four would sit in each quadrant of the base. Each of these four can be a different length, with one going the full distance to the middle, the second meeting the second concentric circle in the base, and the third and fourth only going as far as the third concentric circle.

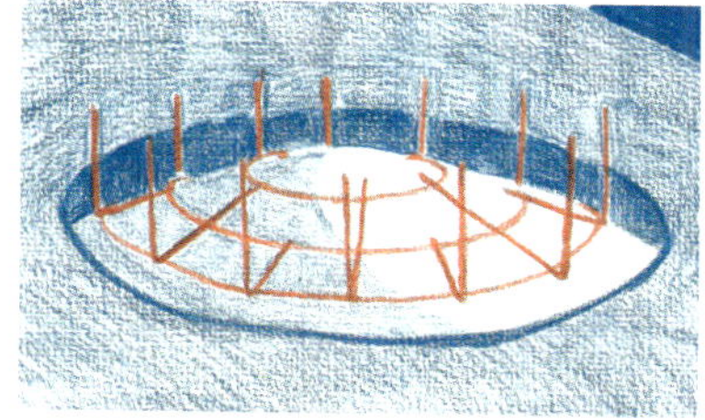
A completed 12 upright foundation rebar design on a domed hole

Add some sand to the pit to get the desired dome shape if wanted, or to fill in any hollows.

Tie all of the rebar together, in situ, using galvanised wire and pliers. Don't fully tighten until you are happy with the position of all pieces.

Place a drainpipe and fasten it to a piece of rebar, making sure it is long enough to protrude from the foundation once poured and located where you intend to slope the foundation. Make sure the pipe and tap, stop valve or other, sit well clear of the cement. Stuff a plastic bag down the pipe to stop any cement getting in there later.

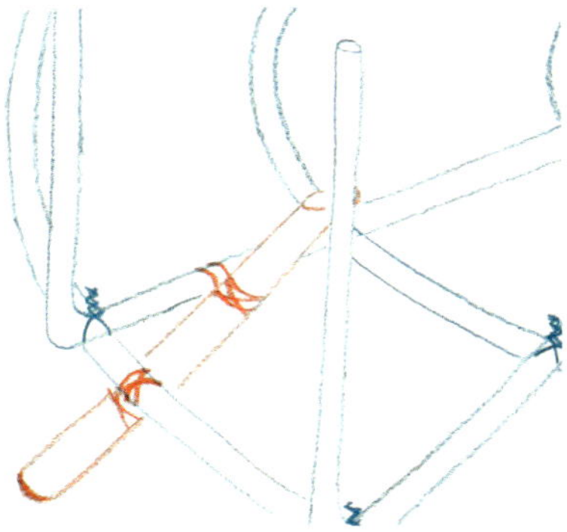

Fastening the drainpipe to the rebar

Using whatever wood and steel you have lying around, hammer shuttering/concrete form work around the hole, about 10cm away

Placing the shuttering ready to pour

from the edge of the steel and at least 10cm deep. This will hold the concrete in position whi e you pour the pad, so make sure it is reasonably robust.

Place bits of broken tiles or small stones under the foundation steel, making sure all of it is hoverirg off the ground and will allow the poured concrete to encase all the steel. Aim for roughly 5cm off the ground uniformly, even when you have made a domed bottom with sand.

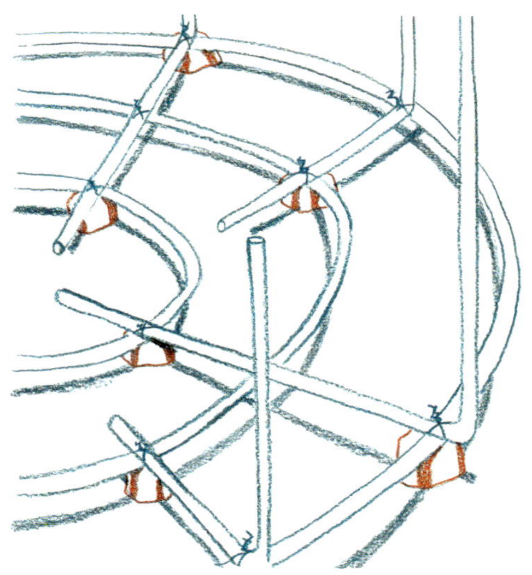

Making sure the rebar is 'hovering' above the ground by at least 3cm

Double check everything! Making particularly sure:

    The steel is off the floor with at least 3cm (preferably 5cm) of clearance underneath.

    The drainpipe is positioned correctly; it should protrude from the concrete by at least 15cm and be stuffed with something to stop concrete going inside.

    The shuttering won't move when concrete is resting against it.

    The upright sections of the right angles are roughly vertical.

Step 5: Pouring

Pour the concrete foundation making sure the depth is around 10cm and that all the steel is well covered by at least 3cm of concrete. We used the following ratio: 1 part cement, 2 parts sand, 1 part gravel. Note: there is so much advice on this topic, so feel free to change this; we were quite cement-heavy and gravel-light compared to other suggestions out there.

Pouring the foundation

Smooth it out with a trowel as you go along, compressing and removing air bubbles, paying special attention to the slope and the drain – a little indentation around the drainpipe can help with cleaning later. Try and get the floor sloping towards the drain – if you mess this up or if you are unhappy with the thickness, you can always add another skim of concrete when the tank is done.

Smoothing the foundation cement, ensuring a depression for, and a slope towards, the drain

Once the foundation is poured, keep it covered and spray it with water to keep it humid as it cures, especially during hot dry weather.

Work can cease here for a while if needs be!

Chapter 9

Making the Cage

Cut all the upright rebar sections to the desired length, remembering to firstly ensure that they begin on or near the concrete slab with the first 40cm or so overlapping with the pieces protruding from the foundation base, and secondly to include the roof in your calculations. We recommend a lid of about 70cm diameter, allowing for easy access with a ladder.

Bend them to create a shallow, sloped roof, making sure they are all the same. Creating an angle between 95° and 105° will give you a roof slope of between 5–15°, which is ideal.

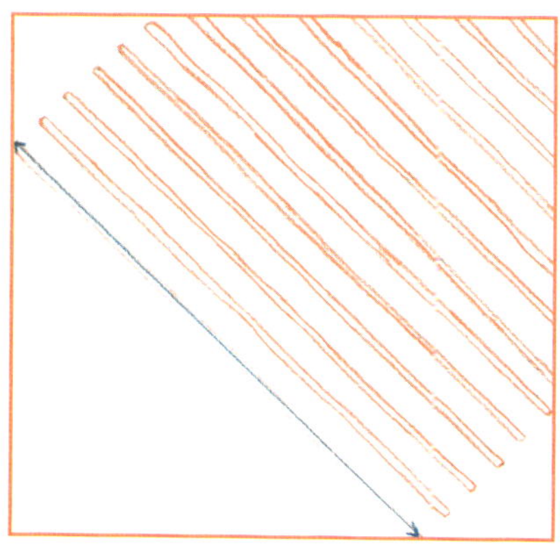
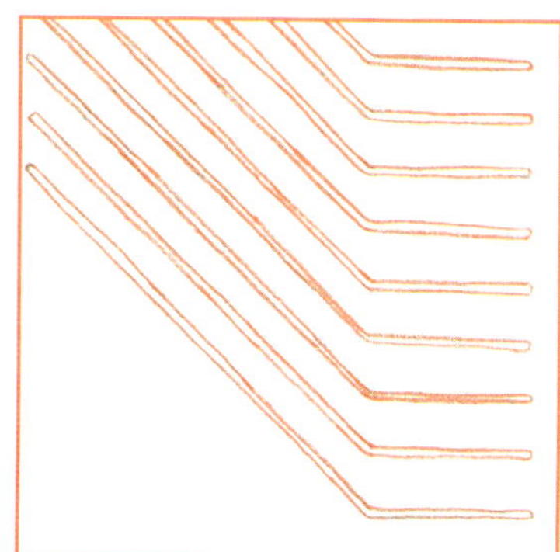

Measuring and bending rebar

Make the hoops, doing your best to make them circular and making sure you have 50cm of overlap. These should be spaced every 30cm up the tank except the first one, which should be 10cm above the foundation. The greatest pressure from the water will be on this junction between wall and floor, so this will give it plenty of strength. If your height doesn't divide neatly into these numbers, err on the side of caution and add an extra hoop.

Attach the uprights to the protruding pieces from the now set foundation with at least three galvanised wire ties. Once attached the uprights can be persuaded towards vertical if needed.

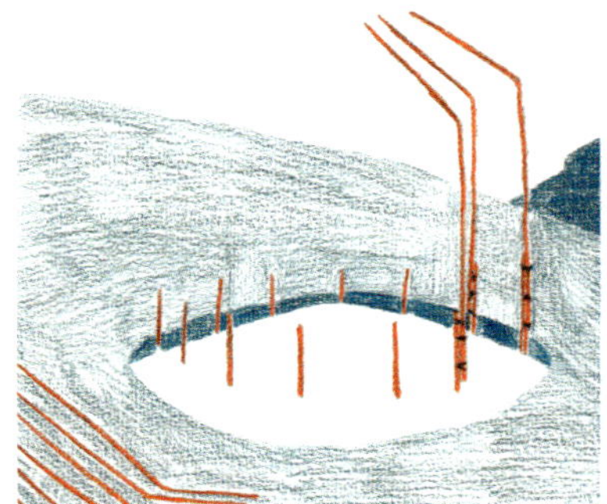

The uprights being attached to the foundation

Slide the first hoop on and fasten it using the wire tying technique outlined on page 27. At this point it will be a tangle of rusty steel swaying about, but worry not, it will get better!

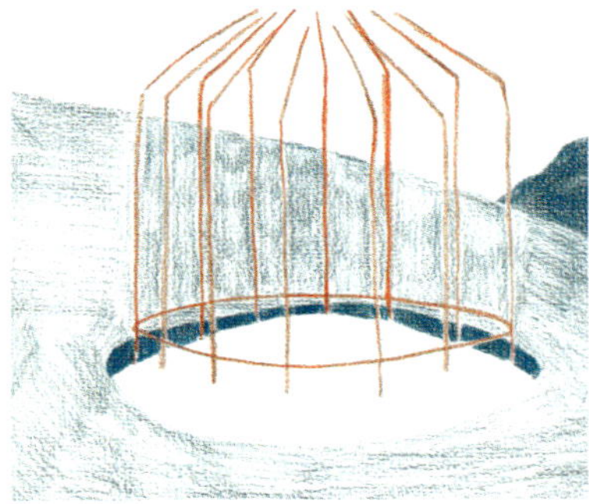

All uprights attached and first hoop on

Slide on all hoops and secure with galvanised wire and pliers, using a hammer or other tool to straighten and adjust as needed, until it is approaching something resembling a cylinder. Often there is a slightly straighter point where the hoops are overlapping; it might be good to either place these bits on a side which won't be visible, or otherwise spread them out around the circle.

Make a hoop/concentric hoops (depending on size) to attach all the roof pieces together, giving you a hole which will be covered by your lid. It is likely that the uprights will be protruding quite a bit into the lid space, but they can either be bent in and cemented in later or trimmed off.

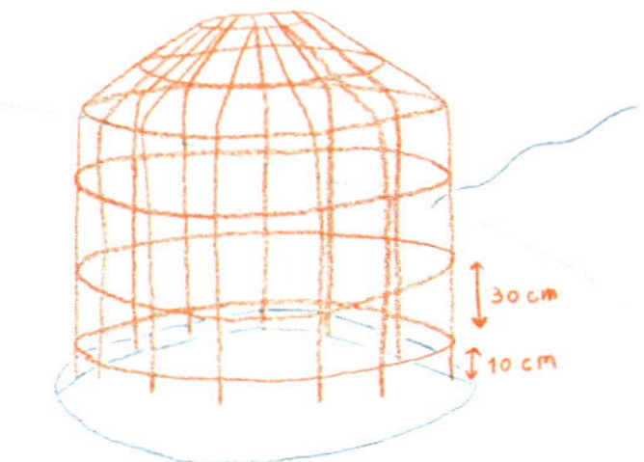

All the hoops attached – note this is a smaller tank design

Hammer or bend in all the sticking out bits of wire. It is worth doing this, otherwise they are a nightmare when plastering and will destroy many gloves!

Attach the chicken wire both inside and out using either wire and pliers or preferably a pair of hog ring pliers. Don't hold back here; time spent

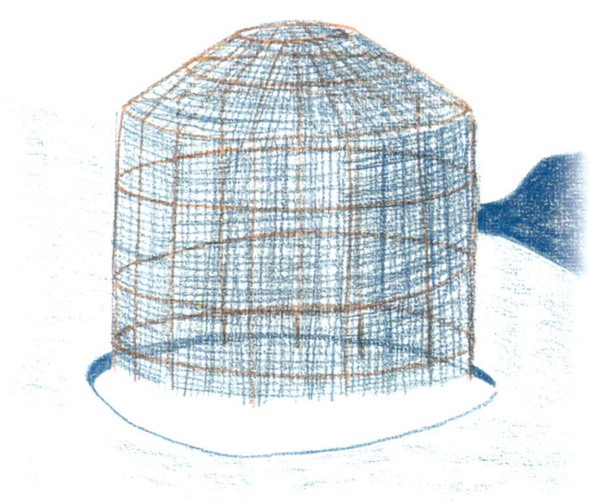

Attaching the chicken wire

First and second layers of chicken wire

on this stage will make plastering much easier. Make the chicken wire tight, without any huge gaps, bulgy bits or loose bits. It helps to have several pairs of hands here to attach the chicken wire snugly.

Position and fasten all taps, pipes, overflows, inlets and outlets using galvanised wire again and twisting it to the best of your ability either against the rebar or the chicken wire. Note: these fittings will feel loose until they are concreted in. Also, cut 'teeth' using a hand saw into anything smooth to help the concrete grip. Just don't cut through!

Putting 'teeth' into the fitting

When positioning your inlet and overflow, be sure to maximise your storage capacity by minimising the vertical distance between the two. If the overflow is 10cm below the inlet, you lose 10cm of tank capacity. The top of the overflow should be cut horizontal; this will again maximise capacity.

Double check everything – walk around the tank a few times checking for anywhere that could do with an extra piece of wire or hog ring, the pipes and taps, and bending in any last pieces of wire so they are flush with the cage.

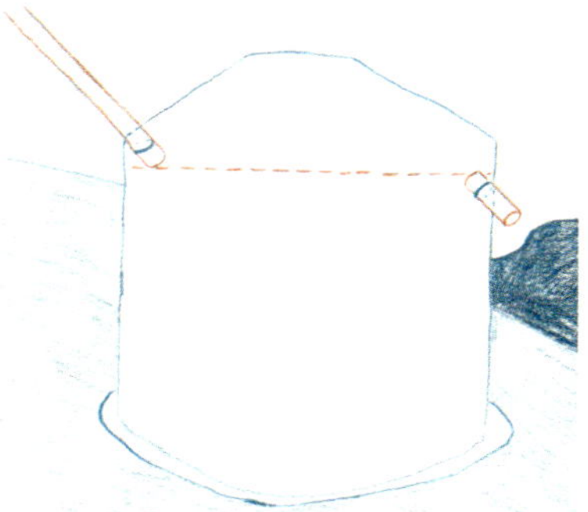

Inlet and overflow details

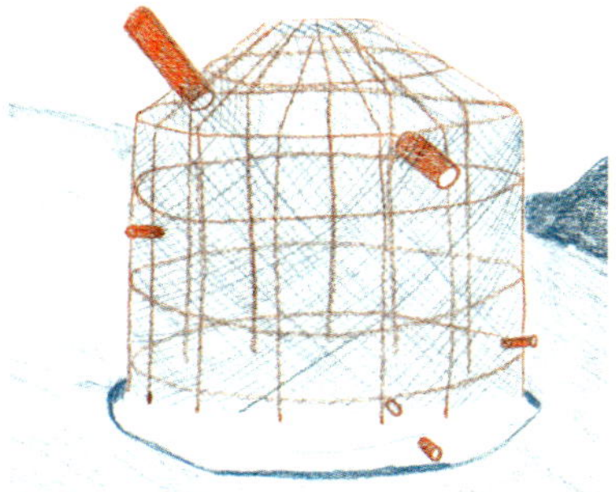

All the fittings attached to the cage

Whisky's Top Tip

Add a tap half way up, or higher perhaps, to give you more ergonomic access when the water is above the level of the tap. Great for filling watering cans and avoiding a bad back, not to mention giving an idea of how full it is.

This is another good time for an interlude if weather or lack of mates/materials demands it!

Chapter 10

Plastering

The mix for the mortar is 1 part cement to 3 parts sand. Water will need to be added as necessary depending on the dryness of the sand, humidity etc. You are aiming for a consistency that allows a dollop of cement to remain stuck to your trowel when turned upside down. Set everything up as ergonomically as possible whether you are doing it by hand or with a mixer.

Cement 1 : Sand 3

You will need one person inside the tank and one person outside; this could be tricky unless you have left a hole in the side as mentioned in the Chapter 7 (page 22) and will require some kind of ladder set-up going through the hole you've left in the top (stay safe!).

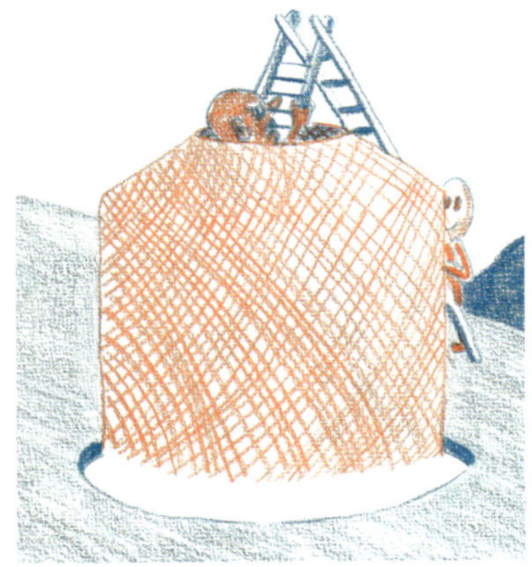

Plastering inside the tank can be tricky; take care

Starting from the bottom, the person on the outside holds two rectangular trowels firmly against the chicken wire, and the person on the inside applies the cement by hand, pushing it through the chicken wire onto the two trowels outside. This will take some getting used to, adjusting your position, hands, firmness and mortar mix, but you'll get the knack – it by no means should look like a finished product after this first coat – in fact, it will look like a mess!

This initial coat provides the structure onto which the two final coats can be carefully applied to create water tightness and an overall thickness of 4-5cm.

Whisky's Top Tip

Applying the cement to the chicken wire requires a knack; the greatest success we had was having one person outside the tank with two large rectangular trowels and one person inside pressing the cement through the wire by hand.

Two pairs of hands absolutely necessary

Pay special attention to all pipes and taps; they need to be carefully plastered in by the most capable person, especially the bits out of your direct eye-line, for these are the most likely places to have leaks. In particular the undersides of taps and pipes positioned towards the bottom of the tank.

Take extra care around the fittings

Getting the first layer on the overhanging roof section can be a nightmare and you will benefit enormously from having either chicken wire with small gaps and/or extra layers to give the concrete mortar something to hang on to. Remember this section doesn't actually come into contact with water so don't panic! In general, with the first layer, apply what you can, allowing excess to fall through. Once what has stuck has gone off, it will provide the strength and structure needed for the next layers.

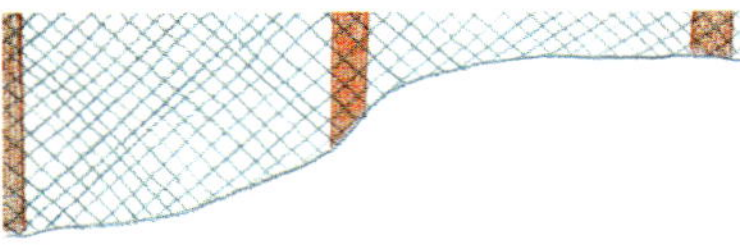

The fillet – any dropped cement can either be reapplied or used to create a 'fillet' at the junction between wall and floor to add further strength both inside and out.

A strengthening fillet at the wall/floor junction

Once this first coat has been applied, keep it out of direct sunshine and spray with water occasionally for two or three days – in a warm dry climate this is especially important, as fast curing cement will weaken your tank.

Keep it sheltered and damp while the concrete cures

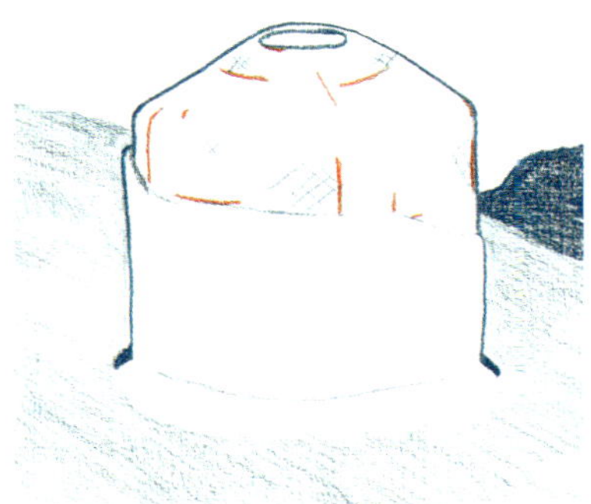

The second coat being applied

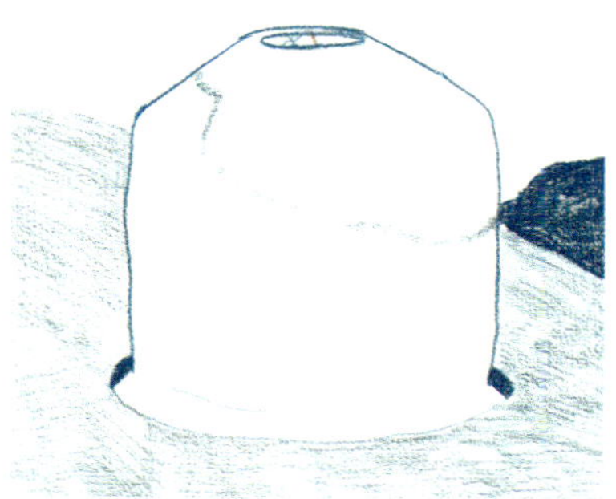

Starting to look tidy now

The finish coats

A day or two later your wobbly steel cage has become a solid structure. For the finish coats we'll start with the outside, but it doesn't really matter. Use the same mortar mix as before, 1 cement to 3 sand.

This is the finish coat, so take your time and use the sprayer to keep the mortar workable while you get it smooth. Take care to push the mortar into crevices and holes in the imperfect first coat and be sure to cover all metal work.

Do as much as you can in one session; this way you avoid cold joints (a join between wet and dry concrete), which can seep a little but will seal up over time with minerals from the water, so don't worry. Also they are not as aesthetically pleasing!

Again, pay special attention to pipes and taps; they need to be solid and water tight. Any inlet pipe of any length might have a fair bit of weight so a small buttress of mortar inside and out will help support this.

Double check the end result

Any spillage can either be reapplied to the wall quickly or used to make a fillet if you haven't already done so, especially on the outside which you probably won't have done on the first, structural coat.

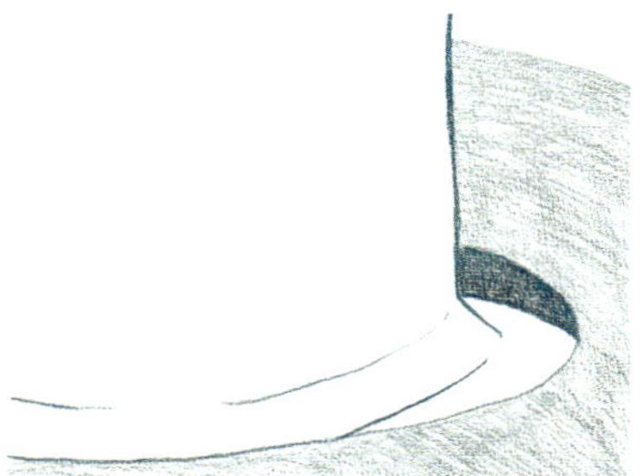

A fillet on the outside uses up spillages and adds strength

Keep all fresh work wet and covered for several days depending on the climate.

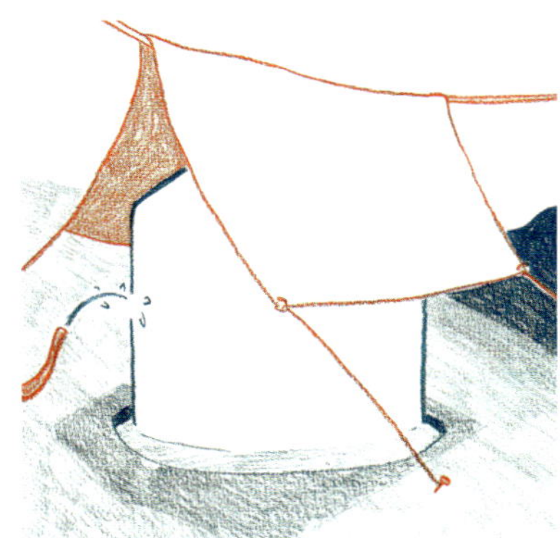

Keep the final coat damp and covered

Optional step:
Make a trough for the drain and lower tap

Whilst not entirely necessary, an external trough can help keep the drainpipe accessible and provide a pleasing aesthetic finish for the drain and lower external tap (if you have one). Also, you probably have leftover bits of rebar, chicken wire and cement which you can use up productively.

We created a square trough with side walls and an open front to frame our tap and drain. We skimmed this with cement and then embedded white pebbles into it for a pleasing finish.

Chapter 11

Making the Lid

On a large, flat area of earth or plywood, mark out the exact dimension of your opening – this should be a circle but probably won't be perfect, so place a sheet of wood across the opening while you are inside the tank and draw around it with a pencil.

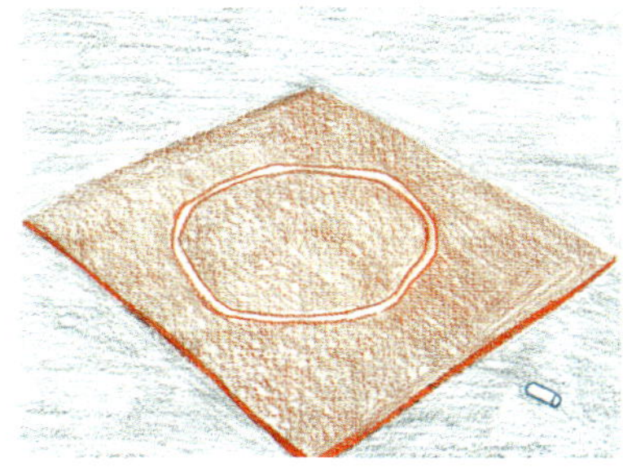

Place small pieces of brick or wood around the pencil mark to create a 5cm-deep edge. Wood may have to be screwed down.

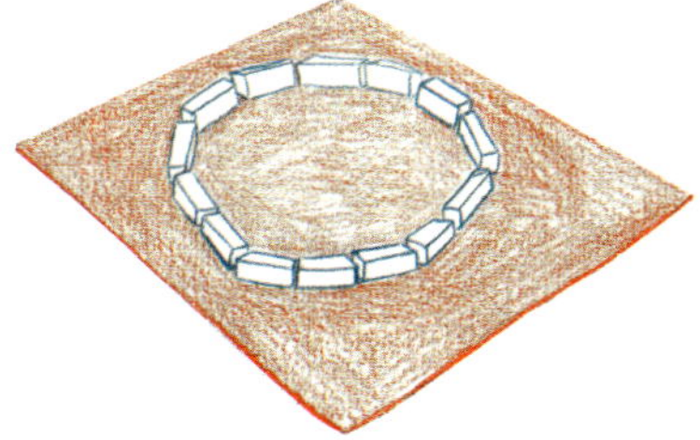

Using some sand, create a gentle dome shape inside the wood or brick perimeter.

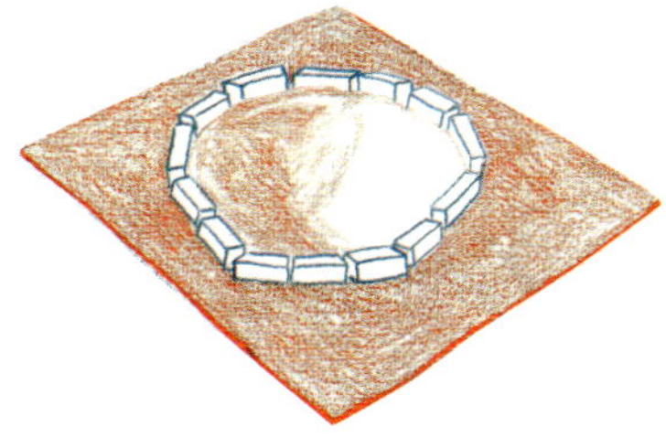

Carefully trowel out some
mortar (the same ratio as the
tank mortar) to cover the whole
area in about 1-2cm.

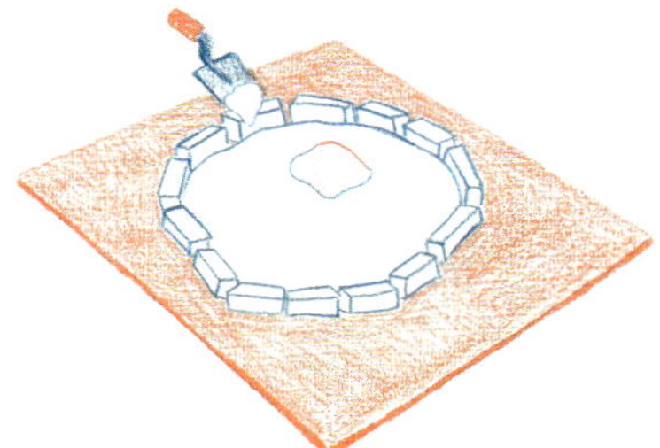

Then gently place a layer of
chicken wire into the wet
concrete over the entire area up
to 1-2cm short of the perimeter.
Also add two handles which will
have plenty of 'key' or surface
area in contact with the next
layer of concrete.

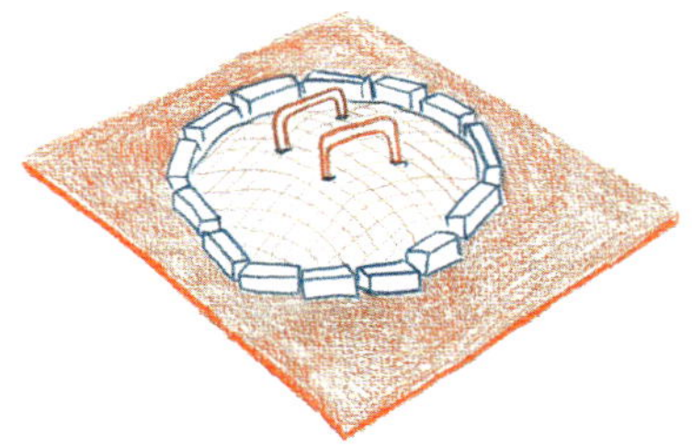

Apply another 2–3cm of mortar
making sure everything is well
covered but also thin; remember
you'll have to lift this onto your
tank.

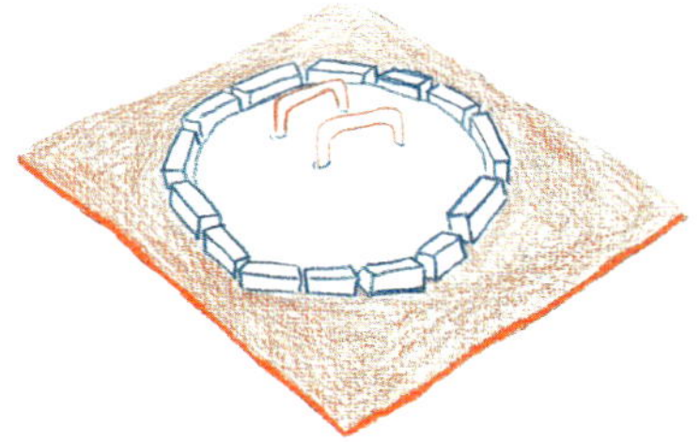

Keep moist and covered for
four days. Meanwhile, place a
small fillet of concrete around
your opening, about 4 x 4cm,
and, using a trowel, carefully
smooth it into a flat rim ready
to receive your lid.

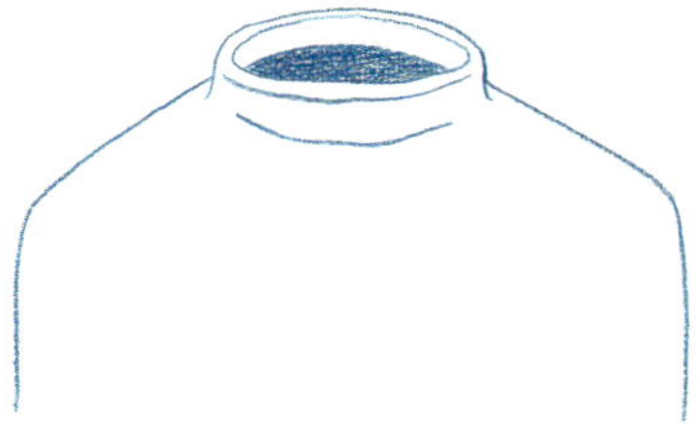

Cement paste to seal your tank

With some cement powder and water, make a paste with roughly the consistency of full-cream milk. Paint the inside with two generous coats and the outside with one coat. Double check below any piping to make sure the plastering is tidy and add more plaster if necessary. Place the lid on top but don't seal it yet.

Filling the tank and checking for leaks

Connect your inlet to the water supply. When the tank fills for the first time check for any leaks. Some damp patches are inevitable, particularly on cold joints, but they will self-seal with time.

If there are any areas of significant seeping, you'll need to empty the tank and repoint with an extra bit of plaster. This happened to us as a result of careless plastering around the wall/floor junction and under a tap situated at the bottom of our 8,000 litre tank.

Tidying up

At this point you can back-fill around the excavation site and landscape around your tank. Backfilling has the added benefit of covering some of the tank with earth and thereby adding the temperature-regulating effect of the ground in both summer and winter. The outside can be painted with two or three coats of limewash; this is non-toxic, reflects a lot of sunshine (it's bright white), is cheap and looks nice!

The concrete lid, previously placed, can now be sealed with a thick cement paste mix. This is easily breakable with the tap of a hammer when you need to get inside and clean it.

Chapter 12

Ongoing Maintenance

Frequency of cleaning will of course depend on your situation. A house surrounded by trees may cause organic debris such as leaves to accumulate in the bottom. A filter or screen as detailed in the Tank Features chapter will help avoid the worst. Maybe after a year it is worth cracking the lid open, draining the tank and giving it a clean; then you can decide whether or not a year is too short an interval or vice versa.

It takes about 45 minutes to clean the tank, having drained it of course, which can take a wh le. It is best to empty and clean your tank when there is some definite rain on the horizon. If a single tank is your only water source, it will be necessary to drain off and keep enough water to last, either in jerry cans or some other storage device, until the tank has ref lled sufficiently. Once there is about 5–10cm of water left in the bottom, close off the drain and use this remaining water to rinse off the walls as you scrub them, then scrub the floor to loosen all the gunk before opening the drain once again. Once the tank is empty, sling a few buckets of clean water around to get rid of any remaining debris and then reseal the tank with some cement paste, close the outlets and hope for rain!

The accumulation of matter on the bottom of the tank is not a problem; it will stay settled there, but if it becomes so deep that it reaches the height of the outlet tap then the water will not flow freely or cleanly. Each tank is unique in this matter, very much depending on its placement and proximity to trees, so doing an investigative clean after a year should help you to determine how long you can go between each clean.

The limewash will also need to be reapplied with two to three coats once every year. It is an easy job and well worth the aesthetic reward.

We live in temperate western Europe, where the summers are hot and the winters are cold. Some freezing has occurred in the winter on the surface of our tanks but caused no damage. It required days and nights of sustained minus temperatures to freeze even just the surface of such a large volume of water. Both the volume and the earth backfilling will have significant temperature moderating effects. It was possible to hear the ice cracking as we drew off water. This could be remedied with an empty plastic bottle floating on the surface of the water; this same technique is used to winterise swimming pools.

Conversely, as well as they tend to hold the warmth during the winter, the same can be said for the summer. Even during consecutive heatwaves where the daily temperature was above 42°C and nights rarely below 20°C, our tanks, which sit in full sunshine, still produced cool clean water throughout a 12-week drought period.

Chapter 13

Further Information

Large Ferro-Cement Water Tank: Design Parameters and Construction Details, UNHCR (2016):
www.unhcr.org/publications/operations/49d089a62/large-ferro-cement-water-tank-design-parameters-construction-details.html

The Humanure Handbook, Joe Jenkins (2013). Jenkins Publishing.

Water Storage, Art Ludwig (2013). ISBN: 0-9643433-6-3 – For a broader study of a variety of water storage solutions and principles, up to a very large scale.

Ferrocement Tank in a Remote Community, Alan Bemo (2014). YouTube – An excellent video outlining the construction process with minimal tools and plenty of hands.

A list of useful books can be found at the end of this brief guide:
www.lboro.ac.uk/orgs/well/resources/technical-briefs/36-ferrocement-water-tanks.pdf

Chapter 14

Case Study

Utilising an off-grid water supply in a temperate climate in the global north

In an effort to give the usage of ferrocement tanks some real world context, this case study outlines how we use the tanks ourselves, how they connect with our off-grid house, why we have the size and number of tanks we do and how we practically use the water to live off.

As mentioned in the Considerations chapter, we have two tanks, one of 3,000 litres and one of 8,000 litres, named Hank the Tank and Frank the Tank, respectively. They provide all of our household needs, which include cooking, three sinks and a shower. We have a dry toilet and no water intensive appliances such as washing machines or dishwashers. This initial reduction in our water usage was key, and prior to thinking about storing water it is perhaps sensible to think about how usage can be lowered.

The drinking water is filtered through a passive carbon filter, a 'Berkey' everything else is used directly from the tanks. The larger tank provides us with cold water and is plumbed into the house using standard plumbing pipe and a 12V, in-line, diaphragm pump inside the house which pressurises the entire system giving us pressurised water when we open any of the taps or the shower. These types of pumps are used in motorhomes and boats, and are thus very energy efficient as well as being little pocket rockets in terms of pressure.

The smaller tank has a pipe which passes through the house via a shut-off valve, gravity feeding an evacuated glass tube solar water heater situated below the level of the water tank. The water heater has a separate 200 litre insulated tank incorporated, which we refill manually as and when we need to. The hot water tank is then plumbed into the house and pressurised with a second 12V diaphragm pump.

All of the plumbing uses standard off-the-shelf compression fittings and HDPE pipe so no welding is necessary, perfect for the novice plumber.

Our tanks have never been less than half full, and this is due to a generous estimation of our water needs. We spent a couple of years living in a caravan whilst we built the straw bale house, a period of time that was immensely educational in terms of our consumption needs for not only water, but also all forms of energy resources. Through collecting our water in jerry cans during this time, we quickly learnt how many litres we used per person per week. For us, this was 60 litres. Now, our needs were incredibly basic at that time; showers consisted of a steel sink of water and a jug, so when planning for the house we allowed ourselves to double our usage to account for new luxuries, i.e. a pumped shower. We used this figure alongside the area of our roof space, coupled with the longest dry period ever experienced in our region to work out a very safe estimate of the amount of water we would need to store. Our caravan usage was around 6,000 litres a year, and we ended up building about 11,000 litres of storage which has been more than enough.

We view our water usage more in terms of 'borrowing'. We capture the rainwater, borrow it for some time in the house, and then, using a reed bed system to filter the borrowed water, direct it to percolate back onto our land going into our swales above the forest garden.

Chapter 15

Conclusion

It has been a privilege to be able to share this technique with others, both in person and through this book. We sincerely hope that by storing this sacred resource, you will gain a new appreciation of our place in this world. In the years ahead these techniques will become crucial as we develop resilience in the face of successive climate upheavals and you may find yourself with curious neighbours wondering why your taps didn't go dry and why you don't have a water bill.

We have been living with this system for three years and its resilience never ceases to astound us. The fact that we have been able to use this very simple system to entirely meet our water needs is remarkable. As mentioned previously, we often inform people that the entire cost of our tank building was less than the quote to join the mains!

Although we use our tanks in France, we truly believe this technique can be employed in most places and climates. Very northern, extremely cold climates are exempt from this, but hotter, drier, more arid areas, which have infrequent rainfall, of which the number is likely to increase, can greatly benefit from storing water in this fashion.

Perhaps unsurprisingly but nevertheless noteworthy is the overnight behavioural change brought about by the undertaking of water stewardship. The rain now brings a smile to our faces, unless it is the end of a long dry summer, in which case it is more like a mad dance. The reawakening of your connection with this sacred natural element will inevitably overflow into other areas of your life, which is exactly the kind of enlightened action needed now more than ever.

Chapter 16

Consultancy

Having embarked on a project to provide yourself with clean, free water storage, you may find yourself in need of technical help. We have often been in this position ourselves during our own journey and recognise the immense help a quick chat can give, to clear up any technical doubts or even just for some reassurance.

You are going to invest a considerable amount of time and effort in this project and when carried out well it will reap enormous dividends. Clean water, on tap, will become something you just don't think about anymore other than with a satisfied smile every time the raindrops fall. With this in mind, we are delighted to offer an online consultancy service, whereby you can contact us with any technical questions and arrange a short video meeting to help clarify things for you.

You can reach us via our website at **www.lesvignesbasses.org** where you can arrange a meeting to suit you.

Creators' Biographies

Felicity Lee

Felicity finds learning and growing the biggest motivations in her life. She revels in discovering knowledge, both old and new, and readily tries to find space to implement these ideas. Felicity's main focus in recent years has been to tread in a lighter manner on the Earth through building and consuming more sustainably, alongside keeping her heart, mind and body open to development, and sharing her beloved practice of yoga.

Daniel Colman

Daniel is passionate about minimising ecological footprint and reconnecting with our deepest essence. Both practically and spiritually, he is fascinated about what follows the current paradigm of extraction, consumption and distraction and hopes to cultivate and disseminate practices that will endure as we navigate the coming tumultuous decades.

Berre Daneels

Berre loves the power of a good story and he loves bringing images to it. He finds creating a whole new world with a few brushstrokes the best thing in the world. Escaping into one of those worlds may seem tempting, so he also tries to make real and meaningful connections with this world and the people in it and in doing so finds stories that are even more wild than he could ever imagine. He hopes that he can bring those things together and tell stories about people in distant worlds that impact this one here.

Enjoyed this book?
You might also like these
from Permanent Publications

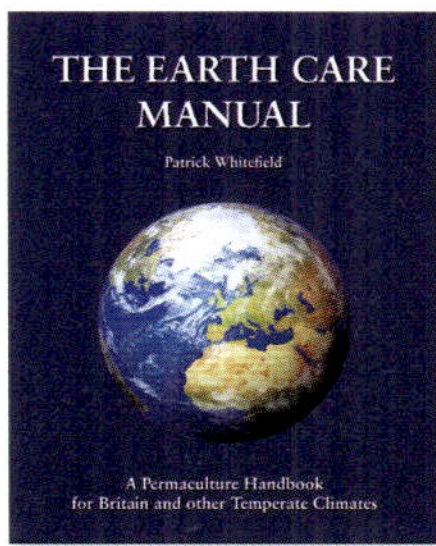

The Earth Care Manual
Patrick Whitefield
£45.00
The definitive design
manual. Show in detail,
how to apply perma-
culture to any situation;
buildings, houses, apart-
ments, gardens, orchards,
farms and woodlands.

**Roundwood
Timber Framing**
Ben Law
£24.95
This detailed, practical,
'how to' book from
permaculture woodsman
is unquestionably a
benchmark for sustainable
green building.

The Woodland House
Ben Law
£16.95
A visual guide to how
Ben built his
outstandingly beautiful
home in his woodland
using straw bales and
roundwood timber from
his own woodland

Our titles cover:

permaculture, home and garden, green building,
food and drink, sustainable technology,
woodlands, community, wellbeing and so much more

Available from all good bookshops and online
retailers, including the publisher's online shop:

https://shop.permaculture.co.uk

with 10% off the RRP on all books

Our books are also available via our American distributor, Chelsea Green:
www.chelseagreen.com/publisher/permanent-publications

Permanent Publications also publishes *Permaculture Magazine*

Enjoyed this book?
Why not subscribe to our magazine

Available as print and digital subscriptions, all with FREE digital access to our complete 30 years of back issues, plus bonus content

Each issue of *Permaculture Magazine* is hand crafted, sharing practical, innovative solutions, money saving ideas and global perspectives from a grassroots movement in over 150 countries

To subscribe visit:

www.permaculture.co.uk

or call 01730 776 582 (+44 1730 776 582)